Tell others what you think of this book!

A quick, honest review
would mean the world to me!

This book is dedicated to my family.

Thank you for your unending patience and support while I worked on "my book".

Big Circle, Little Square: A Book About Shapes

Written by Jojo Carasi
Illustrated by Jojo Carasi

Published by Sweetdreams Monkeybeans Ink

Library of Congress Control Number: 2026915339

Paperback ISBN: 979-8-9965975-0-5
Hardcover ISBN: 979-8-9965975-1-2
E-book ISBN: 979-8-9965975-2-9

Printed in the United States of America
First Edition, 2026

Connect with Jojo's House Books
E-mail: Jojo@JojoCarasi.com
Visit: www.JojoCarasi.com
Facebook: JojosHouseBooks
Instagram: @JojosHouseBooks

A Book About Shapes

Big Circle Little Square

Jojo Carasi

Big circle.

Little circle.

Big circle.

Little circle.

Big circle.

Little circle.

Square.

Big square.

Little square.

Big square.

Little square.

Big square.

Little square.

Crescent.

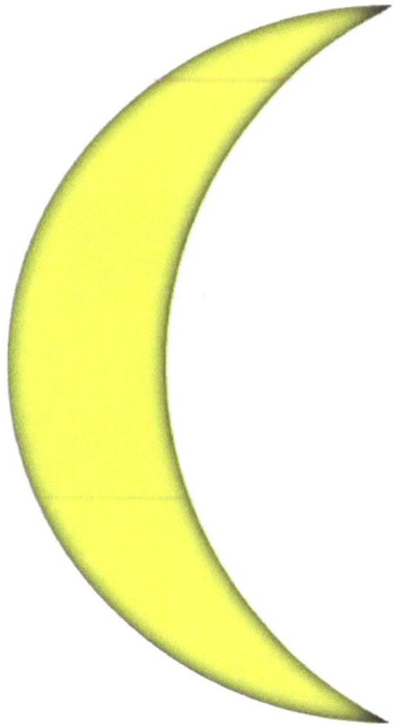

Big crescent.

Little crescent.

Big crescent.

Little crescent.

Big crescent.

Little crescent.

Heart.

Big heart.

Little heart.

Big heart.

Little heart.

Big heart.

Little heart.

Semicircle.

Big semicircle.

Little semicircle.

Big semicircle.

Little semicircle.

Big semicircle.

Little semicircle.

Rectangle.

Big rectangle. Little rectangle.

Big rectangle. Little rectangle.

Big rectangle. Little rectangle.

Triangle.

Big triangle.

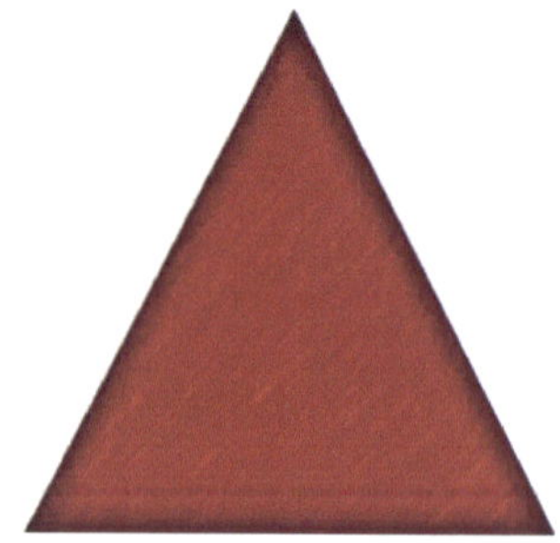

Little triangle.

Big triangle.

Little triangle.

Big triangle.

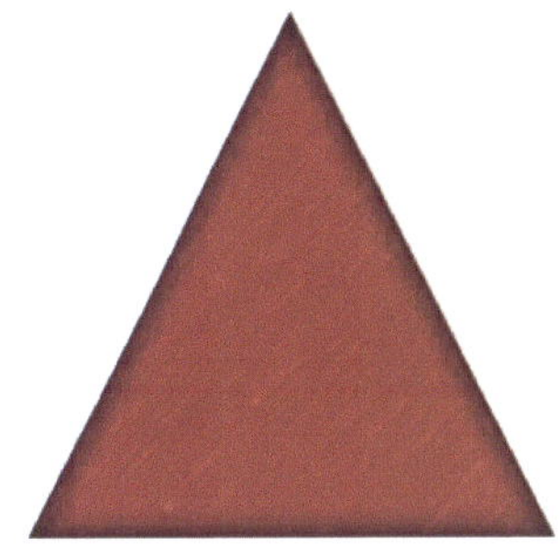

Little triangle.

Oval.

Big oval.

Little oval.

Big oval.

Little oval.

Big oval.

Little oval.

Star.

Big star.

Little star.

Big star.

Little star.

Big star.

Little star.

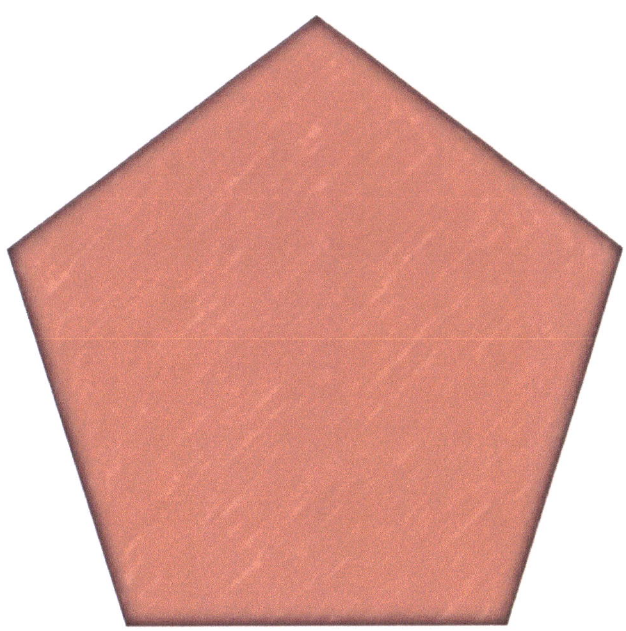

Pentagon.

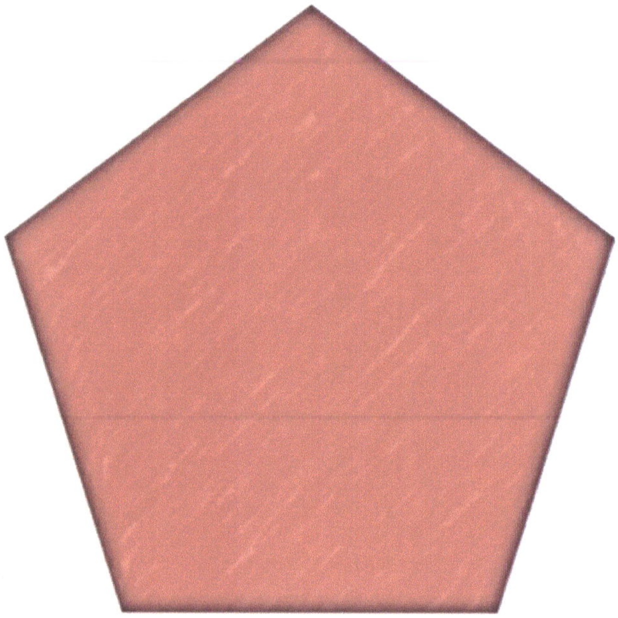

Big pentagon. Little pentagon.

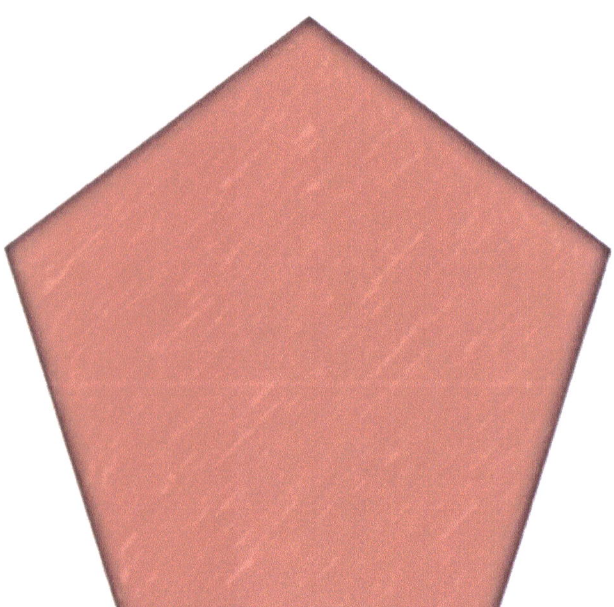

Big pentagon. Little pentagon.

Big pentagon. Little pentagon.

Hexagon.

Big hexagon.

Little hexagon.

Big hexagon.

Little hexagon.

Big hexagon.

Little hexagon.

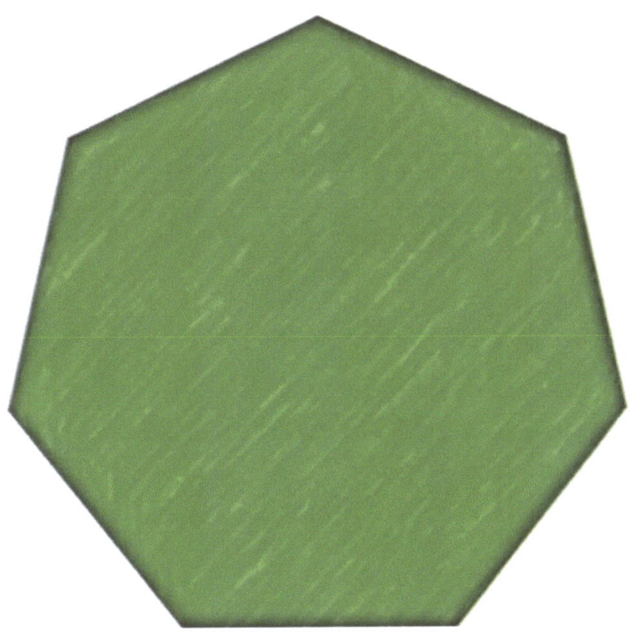

Heptagon.

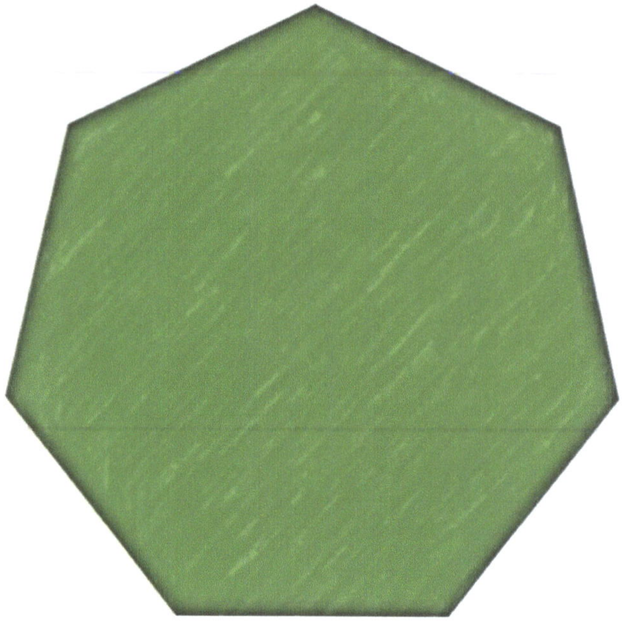

Big heptagon.

Little heptagon.

Big heptagon.

Little heptagon.

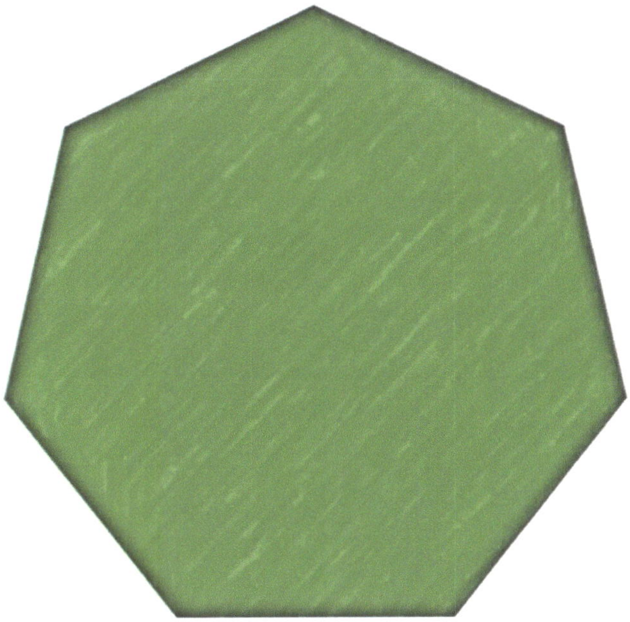

Big heptagon.

Little heptagon.

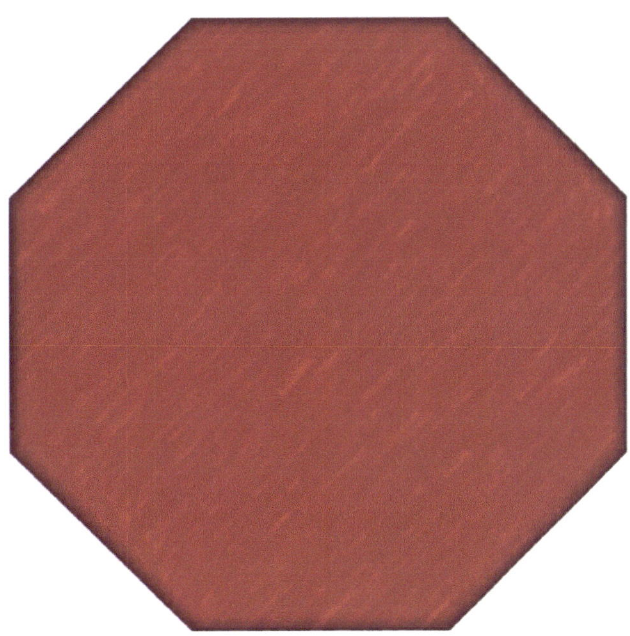

Octagon.

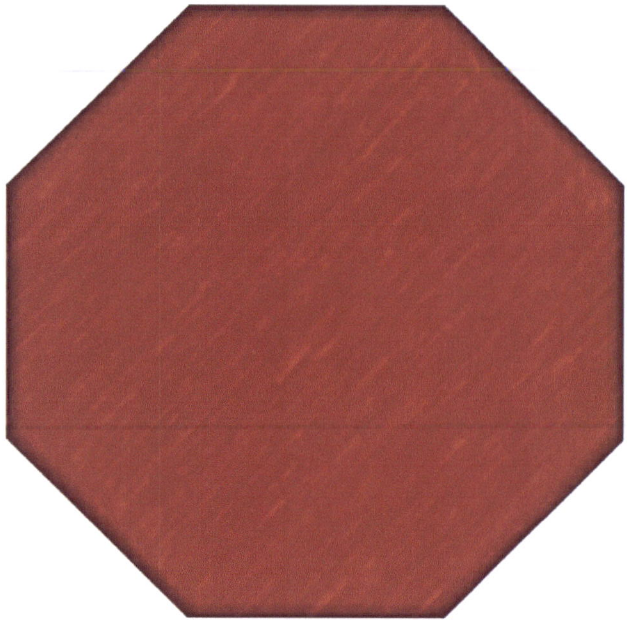

Big octagon. Little octagon.

Big octagon. Little octagon.

Big octagon.

Little octagon.

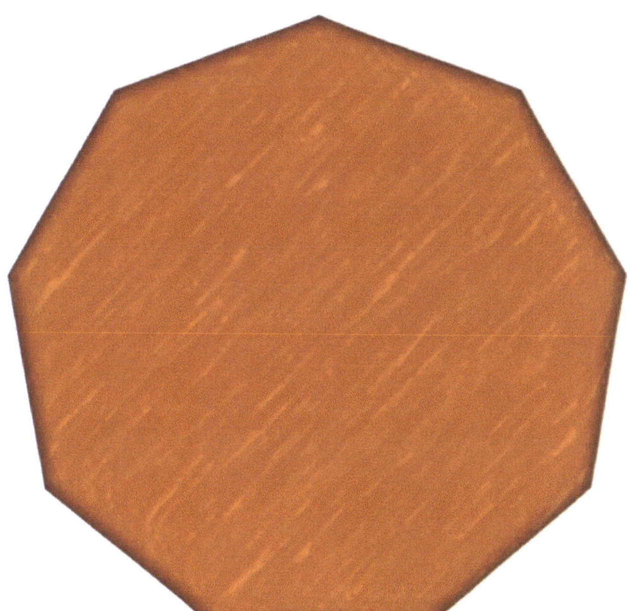

Nonagon.

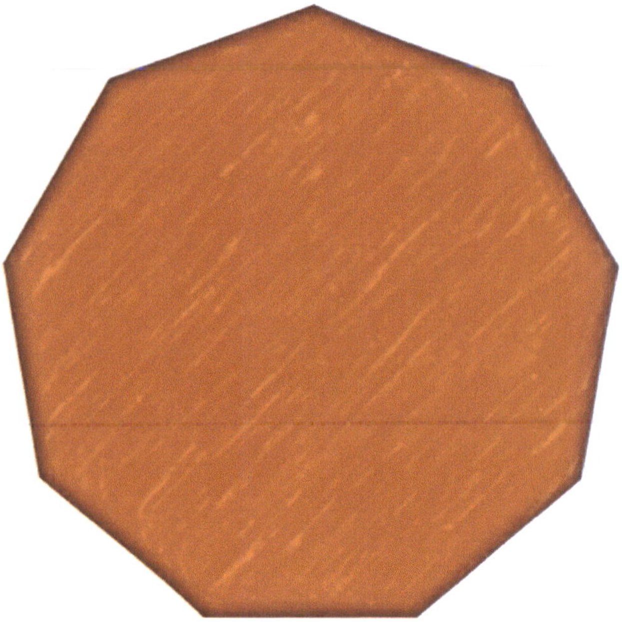

Big nonagon. Little nonagon.

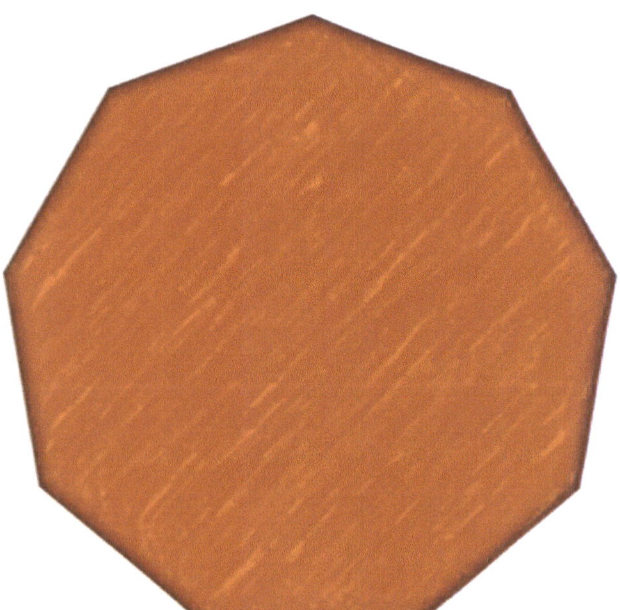

Big nonagon. Little nonagon.

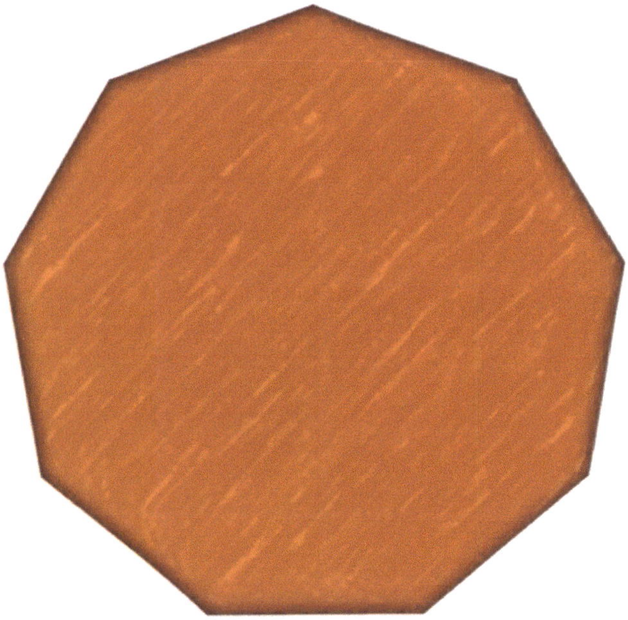

Big nonagon.

Little nonagon.

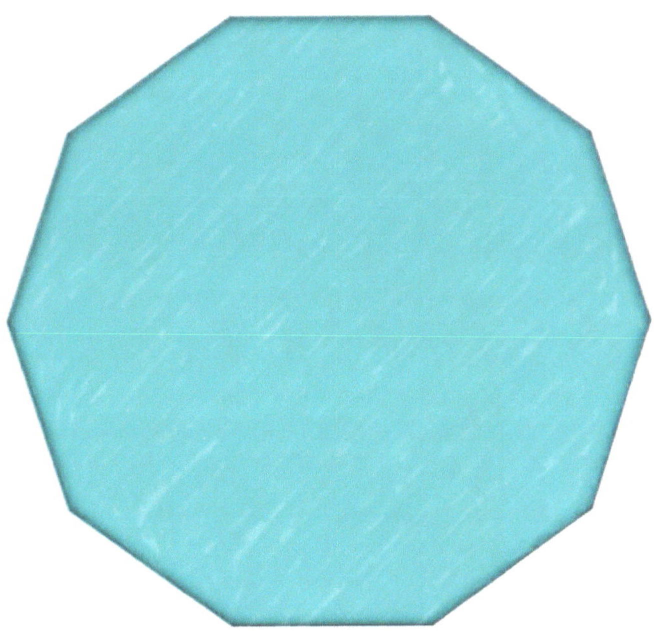

Decagon.

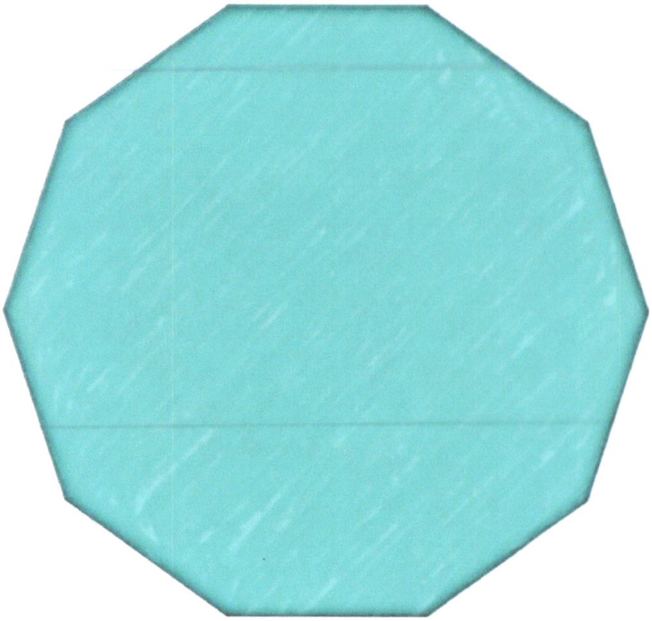

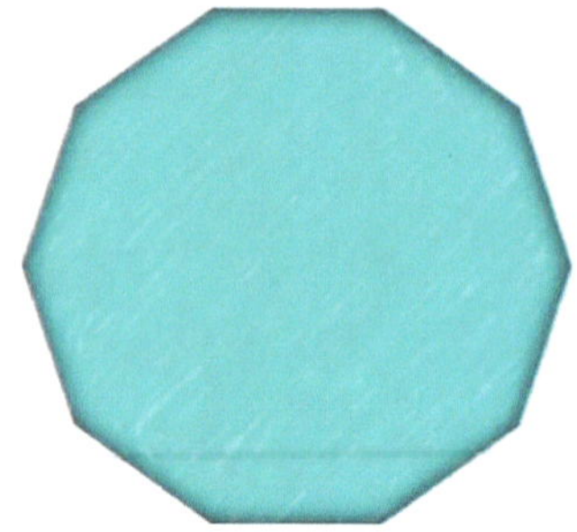

Big decagon.

Little decagon.

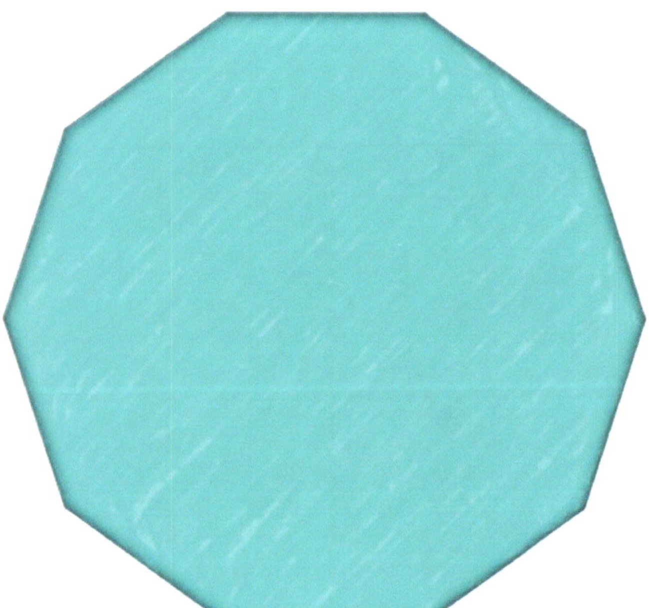

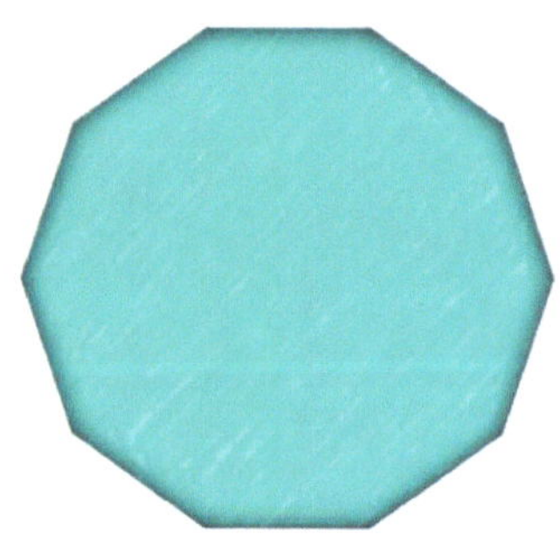

Big decagon.

Little decagon.

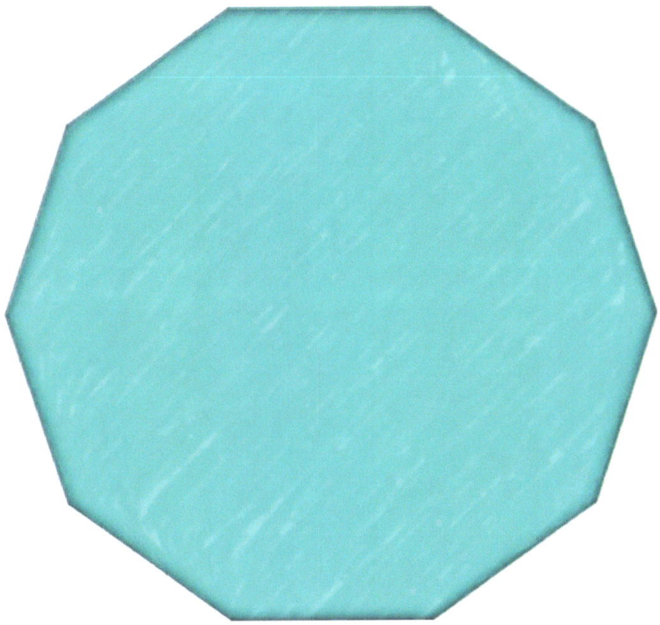

Big decagon.

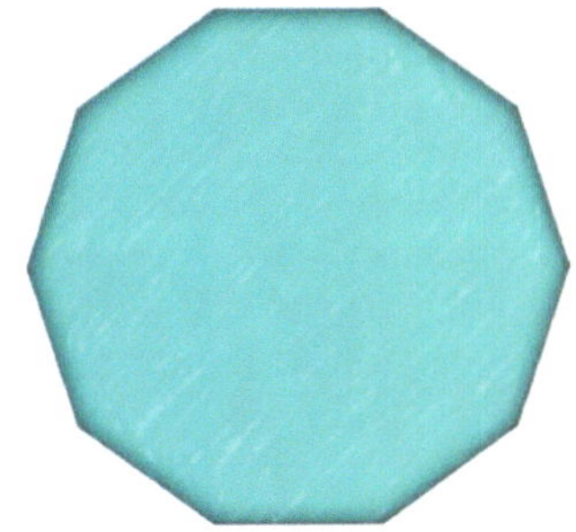

Little decagon.

Arrow.

Big arrow.

Little arrow.

Big arrow.

Little arrow.

Big arrow.

Little arrow.

The End.

Visit

www.JojoCarasi.com

for other books like this and companion printable activities perfect for home and classroom settings.

Instagram: @JojosHouseBooks | Facebook: JojosHouseBooks
Pinterest: @JojosHouseBooks | YouTube: @JojosHouseBooks

At Jojo's House, I've read some books, to the children that were here,
tens of thousands, if I had to count, throughout my whole career.
And dreams of making children's books, have been swirling in my head,
while here, inside at Jojo's House, I've been reading them instead.
But all that changes here today - my journey has begun,
I'm sorting out all of my thoughts to capture one by one.
As finally I put to pen, the ideas in my mind,
I face my fears...I publish them...I hope the world is kind.
This passion I now share with you, you hold it in your hand,
I close my eyes and take the leap, not knowing where I'll land.
After thirty years of reading books, I've turned the page to find,
my wildest dreams are coming true, in a book that I designed.
So read my books, read them again, and share them with your child,
Then search my name, and you will find, the others I've complied.
And if you feel yourself inclined, to leave an honest review,
You will help others find this book - and for that I say thank you!

Tell others what you think of this book!

A quick, honest review would mean the world to me!

www.ingramcontent.com/pod-product-compliance
Lightning Source LLC
LaVergne TN
LVHW071248210926
848486LV00008B/366
* 9 7 9 8 9 9 6 5 9 7 5 1 2 *